Mark the course name, the date, and the lecture or reading topic at the top of your page.

Summarize key points.

Pull out the main ideas or key facts from the right-hand section.

Write condensed versions of notes on the right.

Go for key words or short phrases that communicate the most important information or concepts.

Write potential questions.

Take notes in this section while listening to a lecture or reading text.

Summarize the main ideas here in your own words. Summarize the page of notes asking yourself, "How would I explain this information to someone else?"

Made in the USA
Monee, IL
07 July 2026